Benjamin Schwarz

Comparing the Fire Danger Rating Systems of Canada and the US: FWI versus NFDRS

GRIN Verlag

Bibliografische Information der Deutschen Nationalbibliothek:

Die Deutsche Bibliothek verzeichnet diese Publikation in der Deutschen Nationalbibliografie; detaillierte bibliografische Daten sind im Internet über http://dnb.d-nb.de/ abrufbar.

Imprint:

Druck und Bindung: Books on Demand GmbH, Norderstedt Germany
ISBN: 978-3-656-73243-3

This book at GRIN:

http://www.grin.com/en/e-book/280035/comparing-the-fire-danger-rating-systems-of-canada-and-the-us-fwi-versus

Benjamin Schwarz
7th April, 2014

Comparing the Fire Danger Rating Systems of Canada and the US: FWI versus NFDRS

Abstract

This paper describes and compares conceptually the Fire Weather Index (FWI) system of Canada and the National Fire Danger Rating System (NFDRS) of the USA. The relatively simple FWI system rates fire danger for all Canada and is based on empirical field-data derived from a single fuel type. The laboratory-based NFDRS, in contrast, allows more specification for distinct fire danger areas and models fuel moisture more abstract and in different classifications. Relative strengths and weaknesses with regard to the vegetative conditions in the particular country are discussed. The use of empirical data and the good and simple representation of soil moisture are the major strengths of the FWI system. The NFDRS wins through its possibility to model specifically a distinct fire danger area and through the consideration of live fuel moisture content. The conclusion of this paper is that both systems can benefit from each other. A combination of the simplicity of the FWI and specialization on a distinct area through the site descriptors similar to the ones of the NFDRS is proposed.

1. Introduction

Deeming et al. (1972) defined Fire danger as the "resultant descriptor of the combination of constant and variable factors which affect the initiation, spread and difficulty of control of wildfires on an area" . This descriptor helps fire managers to make decisions and pre-allocate resources so that fires can be fought quickly and efficiently (Schlobohm & Brian 2002).

In North America there are two systems that rate fire danger: the Canadian Fire Weather Index (FWI) and the American National Fire Danger Rating System (NFDRS). Van Wagner (1975) already compared both system structures almost 40years ago. Since then especially the NFDRS changed significantly. In this paper, I will describe and compare the current models regarding model basis, structure and outputs. In the end strengths and weaknesses of each model will be discussed.

2. Description of the Canadian Forest Fire Weather Index (FWI)

The FWI system rates the potential of fire in a single standard fuel type (jack-/lodgepole pine) with six numerical values (Wotton et al. 2009). It is basically a set of indices that represent fire danger at the midafternoon peak at 1600 hours LST (van Wagner 1987). The FWI estimates fire behavior through three moisture content classes and the effects of wind (van Wagner 1987). The system is mostly based on data that is derived from field experiments and observation (van Wagner 1987). Laboratory experiment data was only used to complement the field data (van Wagner 1987). Key assumptions of the model are that fuels and their moisture are homogenous and that wind is only blowing from one single direction without any interaction with topography (van Wagner 1987). Also, every rainfall is considered as overnight rain that dries out during the day (van Wagner 1987). The system components are determined by daily weather inputs like air-temperature, relative humidity, open wind speed at ten meters above ground and precipitation measured at noon LST (van Wagner 1987). The structure of the FWI system is relatively simple and easy to understand as Figure 1 shows.

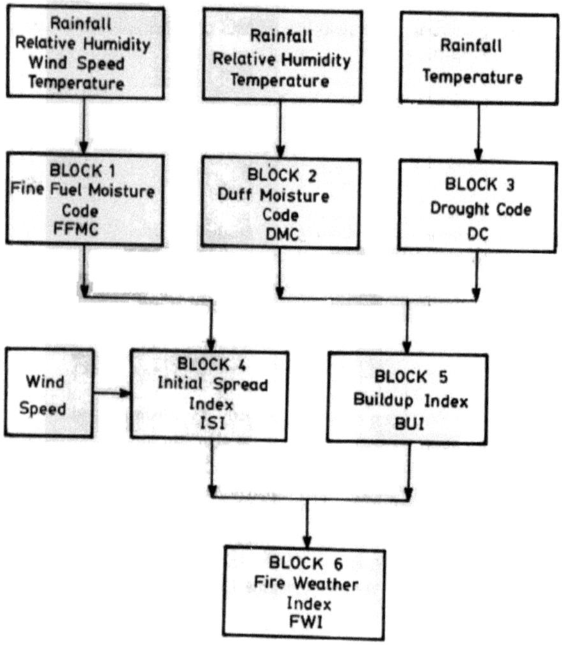

Figure 1 Structure of the FWI system (van Wagner 1987)

The three fuel moisture codes make the basis for the model. They can be seen as bookkeeping systems that track the fuel moisture just by the weather inputs. All the moisture codes go up in value when the actual moisture decreases which implies that the Fire Danger is going up (van Wagner 1987). From the moisture codes the actual moisture content can be directly calculated (van Wagner 1987). The **Fine**

Fuel Moisture Code (FFMC) represents litter and other cured fine fuels on the forest floor (van Wagner 1987). All the measured weather inputs (rainfall, relative humidity, wind speed, temperature) are assumed to have an impact on this code (van Wagner 1987). The FFMC layer has a nominal depth of 1.2 cm and a nominal fuel load of 0.25kg/m² (van Wagner 1987). It is the only moisture code in the model that has varying equilibrium moisture content (van Wagner 1987). The **Duff Moisture Code (DMC)** represents the layer below the FFMC layer with loosely compacted, decomposing organic matter (van Wagner 1987). It is 7cm deep, has a fuel load of 5kg/m² and is affected by rainfall, relative humidity and temperature (van Wagner 1987). The equilibrium moisture content is assumed to be 20% constantly (van Wagner 1987). The **Drought Code (DC)** represents the deep layer of compact organic matter underneath the DMC layer (van Wagner 1987). As the name says, it is an indicator of drought and it only gets determined by rainfall and temperature (van Wagner 1987). It is 18cm deep with a fuel load of 25kg/m² (van Wagner 1987). The equilibrium moisture content is assumed to be constant, too, at 0% (van Wagner 1987).

The next set of indices encompasses the **Initial Spread Index (ISI)** and the **Build-up Index (BUI)**. The ISI is a function of wind and FFMC and indicates the fire spread-potential in a region with the assumption that there is no variability in the quantity of available fuel (van Wagner 1987). The BUI is calculated from the values of DMC and DC and represents the total availability of fuel for a spreading fire (van Wagner 1987). The higher the DMC or DC is the more fuel is available for combustion. Both, ISI and BUI, go into the final calculation of fire danger represented by the **Fire Weather Index (FWI)**. It indicates the potential intensity of a spreading fire under current conditions for the mid-afternoon peak (van Wagner 1987). The FWI is like all the other values of the system unitless. The FWI value gets classified into different fire danger classes, usually ranging from low, moderate, high to extreme. This way fire danger can be easily communicated to the local population though the classification varies in Canada from Province to Province.

3. Structure of the National Fire Danger Rating System (NFDRS)

The NFDRS was intentionally designed to be adaptable to the particular needs of local fire managers, applicable anywhere in the country, all scientifically based and reasonably priced (Schlobohm & Brian 2002). The system models represent principles of the physics of combustion (Schlobohm & Brian 2002). Coefficients and constants were developed in the laboratory and reflect the relationship between various fuels, weather conditions, topography and risk conditions (Schlobohm & Brian 2002). Factors like ignition temperature of woody material and the moisture of extinction of certain fuel types were all laboratory-determined (Schlobohm & Brian 2002).

Key assumptions in the NFDRS are that first, the outputs relate only to the potential of an initiating fire (Schlobohm & Brian 2002). That means that fuels are continuous, the slope is uniform and the fire spreads without spotting or crowning (Schlobohm & Brian 2002). Second, the outputs only address fire activity from a containment standpoint, so full extinguishment is not considered (Schlobohm & Brian 2002). Third, outputs are relative and not absolute (Schlobohm & Brian 2002). Also they are linear, so a doubling of a component or index means a doubling of the associated activity of that element (Schlobohm & Brian 2002). Fourth and last, the ratings represent near worst-case-scenarios with the assumption that the inputs were measured at exposed location at or near the peak of the usual fire season (Schlobohm & Brian 2002).

The structure of the NFDRS is presented in Figure 2 and is apparently more complex than the FWI system.

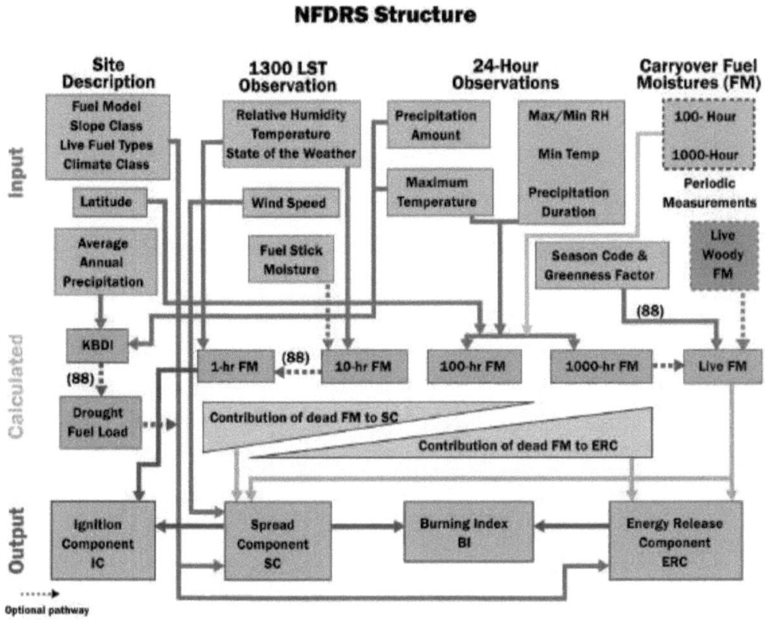

Figure 2 Structure of the NFDRS (Schlobohm & Brian 2002 adapted from Deeming et al. 1977; Burgan 1988)

The user has to provide site descriptors. It begins with the Fire Danger Rating Area where the user geographically defines a distinct and ideally uniform fire danger area with unique fire related decisions (Schlobohm & Brian 2002). Uniform means that fuels, weather and topographic features are rather homogenous (Schlobohm & Brian 2002). Next, the user chooses the **Fuel Model** that mirrors best the type of fuel in the fire danger area (Schlobohm & Brian 2002). These Fuel Models reflect the volume, size, weight, type, depth, surface-to-volume ratio etc. of the fuel bed and determine in how far dead or live plant material contributes to local fire danger (Schlobohm & Brian 2002). Including subclasses there are 20 fuel models in total (Schlobohm & Brian 2002). The six major ones are (1) Lichen and mosses, (2) marsh grasses and reeds, (3) grasses and forbs, (4) brush, shrub and tree reproduction, (5) trees and (6) slash (Schlobohm & Brian 2002). These fuel beds are not an exhaustive representation of the fuel beds in the US but are supposed to be a reasonable match (Schlobohm & Brian 2002). The next descriptor to choose is the **Slope Class** which is very significant in the whole model (Schlobohm & Brian 2002). There are five of them: 0-25%, 26-40%, 44-55%, 56-75% and >75% (Schlobohm & Brian 2002). The NFDRS respects differences in the seasonal growth of grass which has an effect on fire danger. Therefore the **Live Fuel Type** (grass type) can be chosen as well (Schlobohm & Brian 2002). At last, the user selects the **Climate Class** (Schlobohm & Brian 2002). The Climate Class considers differences in seasonal and growth-related patterns (Schlobohm & Brian 2002). There are four to choose from: arid/semi-arid desert or steppe, semi-humid climate with a deficiency in moisture over the summertime, semi-humid with enough moisture to sustain plant-growth even over the summertime and wet coastal areas (Schlobohm & Brian 2002).

Annual precipitation is not necessarily a part of the user-controlled site descriptors but it is a necessary input for the Keetch-Byram Drought Index (KBDI) which addresses the effects of long-term drying. This input should be from local climatological data and (Schlobohm & Brian 2002).The KBDI is obligatory for the 1988-version of the NFDRS (Schlobohm & Brian 2002).

The most important data inputs into the NFDRS are weather inputs either measured at 1300hours LST or as a 24-hour observation. They are the most significant predictors in the changes of fire danger from day to day (Schlobohm & Brian 2002). A continuous input of air-temperature, relative humidity (including 24 hours maximum and minimum), rain duration, open wind speed at 6.1m height and state of the weather (precipitation, cloud cover) is required to keep up the accuracy of the system and to account for daily effects (Schlobohm & Brian 2002). Apart from weather data, there are several other parameters that influence the outputs of the NFDRS. The **State of Herbaceous Vegetation** is a parameter that takes into account when the growing season started (Schlobohm & Brian 2002). This has to be observed and provided by the user (Schlobohm & Brian 2002). Also the **events of killing frosts** and the **Shrub Type Code** (deciduous or evergreen) have to be observed and entered (Schlobohm & Brian 2002). The outputs of the NFDRS can be displayed in groups or classes (Schlobohm & Brian 2002). That's why the user has to deal with the parameter "**Staffing Index and Display Class Break Points**". This parameter wants to know which indicators or components of the model shall be considered to base the Staffing or other outputs on (Schlobohm & Brian 2002). Afterwards, one can also determine the number of class breakpoints or classes (Schlobohm & Brian 2002). This is done because a specific output of the system can be directly tied to a particular action (Schlobohm & Brian 2002). The number of classes reflects the number of different actions that are defined by the fire manager or agency (Schlobohm & Brian 2002). The term "staffing" will be explained later in the paper. **Measured Woody Fuel Moisture** values from local samples can be integrated into the model every month in order to calibrate it towards the actual moisture content of small branches of wood and live foliage in the fire danger rating area (Schlobohm & Brian 2002). Intentionally, this value was supposed to only deliver a point of comparison to the modelled Woody Fuel Moisture (Schlobohm & Brian 2002). A fire manager would take the sample values and compare them to the modelled values (Schlobohm & Brian 2002). With his experience he could then better rate the actual fire danger in his area and make appropriate decisions (Schlobohm & Brian 2002). **Season Codes** and **Greeness Factors** are parameters that are only required for the 1988-fuel models and refer to herbaceous and shrub vegetation (Schlobohm & Brian 2002). The Season Codes correspond to the season of the year and the equivalent stages of plant development while the Greeness Factors represent the state of the plants on a scale from cured to full flush (Schlobohm & Brian 2002). The **Keetch-Byram Drought Index** (KBDI) is actually a stand-alone index that measures the effects of seasonal droughts on fire potential; however, it is required for the 1988-NFDRS version because it determines the drought fuel load in the 1988-fuel models (Schlobohm & Brian 2002). The KBDI estimates how much precipitation is needed to saturate the 8 uppermost inches of the soil (20.32cm) (Schlobohm & Brian 2002).

The final outputs of the NFDRS are fed by intermediate outputs which will be described in the following. The intermediate outputs are what the next day's calculations are based on. They are fuel moisture values (in percent of the oven dry weight of the fuel) of live and dead fuel types (Schlobohm & Brian 2002). The first one is the **Herbaceous Fuel Moisture** which is an estimate of the moisture content (mc) of the live herbaceous vegetation that ranges from 0 to 250% (Schlobohm & Brian 2002). The accuracy in choosing the right Fuel Type and Climate Class is crucial for receiving the correct outputs (Schlobohm & Brian 2002). The **Woody Fuel Moisture** gives out the mc of live woody vegetation like shrubs, small stems, branches and foliage (Schlobohm & Brian 2002). Here again it is important to choose the correct Climate Class since the NFDRS assumes different default starting values for different Climate Classes (Schlobohm & Brian 2002). **Dead Fuel Moisture** predicts the mc of dead organic fuels based on roundwood fuel moisture parameters (Schlobohm & Brian 2002). This part is classified by time-lags and divided into four time-lag classes. Time-lag, in a simple sense, can be seen as the "half-life" of a fuel on the way to reach equilibrium moisture content. It is the time necessary for a (dead) fuel particle to "lose approximately 63% of the difference between its initial mc and the equilibrium mc in its current environment" (Schlobohm & Brian 2002). The ***1-hour Fuel Moisture Content*** estimates the mc of dead fuels less than 0.635cm in diameter which is supposed to represent the uppermost layer of litter (Schlobohm & Brian 2002). Values range from 1 to 80% mc (Schlobohm & Brian 2002). The ***10-hour Fuel Moisture Content*** is an estimate for roundwood fuels 0.635 to 2.54cm in diameter (Schlobohm & Brian 2002). These fuels are supposed to be 1.9cm below the surface and modeled through factors like length of day (determined by latitude and calendar day), cloud cover, temperature, relative humidity and for the 2002 NFDRS-version even solar radiation (Schlobohm & Brian 2002). This layer has a modeled mc of 1 to 60%. The ***100-hours Fuel Moisture Content*** accounts for dead fuels 2.54 to 7.62cm thick which lie

1.9 to 10.16cm below the surface (Schlobohm & Brian 2002). The determining factors are length of day, maximum and minimum temperature of the day, relative humidity and the duration of precipitation in the last 24 hours (Schlobohm & Brian 2002). Values go from 1 to 50% (Schlobohm & Brian 2002). The last one is the ***1000-hours Fuel Moisture Content***. It represents the mc of fuels from 7.62 to 20.32cm in diameter that lie more than 10.16cm below the surface (Schlobohm & Brian 2002). The 1000-hours Fuel Moisture content represents an average calculation over the last 7 days (Schlobohm & Brian 2002). Factors include length of day, the daily average temperature, the extremes in relative humidity and the 24 hours precipitation value for a 7 day period (Schlobohm & Brian 2002). These fuels are modeled to have a mc between 1 and 40% (Schlobohm & Brian 2002). There is an output outstanding from the actual dead fuel classes. It is the X-1000-hours Fuel Moisture Content which is derived from the 1000-hours Fuel Moisture Content (Schlobohm & Brian 2002). It represents the live fuel recovery value and goes into the calculation of herbaceous fuel moisture (Schlobohm & Brian 2002).

The final outputs of the NFDRS are three components and one index. The first component is called **Ignition Component (IC)**. The IC rates the probability that a firebrand will require a fire suppression action (Schlobohm & Brian 2002). This gives a fire manager a direct recommendation for mobilizing and allocating resources (Schlobohm & Brian 2002). The IC is fed by the **Spread Component (SC)** since only a spreading fire requires counter-actions. The SC is influenced by wind speed, slope and fire fuel mc (Schlobohm & Brian 2002). The SC indicates how fast a fire will move forward (Rate of Spread) under ideal conditions (Schlobohm & Brian 2002). The **Energy Release Component (ERC)** estimates the potential heat release per unit area in the flaming front of the head of the fire and accounts for the contribution of dead and live fuels to the potential fire intensity (Schlobohm & Brian 2002). It also reflects drought conditions because the dry out of fuels adds up in the model (Schlobohm & Brian 2002). The ERC is fed by the 1000-hours Fuel Moisture Content (Schlobohm & Brian 2002). The **Burning Index (BI)** represents the "difficulty of control" of a fire related to flame length (Schlobohm & Brian 2002). The

BI is determined by two factors: the Rate of Spread of the potential fire (SC) and the energy released on the way (ERC) (Schlobohm & Brian 2002).

4. Comparison and Discussion

A comparison between the fire danger rating systems of two of the largest countries in the world is only thorough if the differences in vegetation are also acknowledged. The USA lies south of Canada. According to the latitudinal diversity gradient species richness increases towards the equator, so the diversity of species and vegetation types are higher in the USA than in Canada (see Brown & Lomolino 1998). For a Canadian system it is therefore easy to focus on a single fuel type for fire danger predictions as the modelled fuel type (jack-/ lodgepole pine) is widespread throughout the country (van Wagner 1987). A US system has to be locally or regionally adaptable because differences in vegetation and fuels are significant thinking of the conifer fuels in Montana for example and the subtropical forest types of Florida. This fact has to be kept in mind when comparing the two systems.

NFDRS and FWI share some similarities: both of them track the moisture content of different fuel classes daily and therefore require a continuous input of weather data, fuels dry exponentially, the predicted fire behavior accounts for the effect of wind, they have one index to represent a linear Rate of Spread (ISI, SC) and one index to predict the amount of fuel consumed (BUI, ERC) (van Wagner 1975). Also the final outputs are almost similar functions that describe the same physical quantity: line-fire-intensity (van Wagner 1975). The NFDRS represents worst-case scenarios (Schlobohm & Brian 2002). This way fire managers act rather on the safe side and will likely allocate sufficient resources. In the way the FWI moisture codes are modelled they represent worst-case scenarios as well because the rain is supposed to have fallen over night so the fuels can maximally dry out during the day which increases the predicted fire danger.

Both systems have the purpose to rate fire danger so most similarities lie in the outputs of the systems. Yet, there are significant differences in the way of getting to these outputs regarding model basis and used inputs.

The empirical approach of the FWI is a better reflection of real-life conditions as they appear in the field. This is the first significant strength compared to the NFDRS. The laboratory approach of the NFDRS is still necessary since this allows the model to be applicable in several fuel types. While the FWI simplifies the system and restricts itself to field data of one fuel type the NFDRS has to complicate the system. A more complicated system encourages more system bugs and mistakes and requires more training to work with it.

The NFDRS provides through the inclusion of site-descriptors and the possibilities for "fine-tuning" more specialized outputs for local fire danger rating areas. This is an advantage that the FWI system lacks. The choice between the 20 fuel type models, the five slope classes and four climate classes in the NFDRS is not too big but still it is seen as adequate (Schlobohm & Brian 2002) and it provides a certain degree of specification. The FWI output gets only classified and interpreted differently from Province to Province. So the accounting for differences of fire danger only happens on a very coarse scale.

When it comes to the moisture codes in the FWI system the FFMC is the only code with variable equilibrium moisture content (emc). Since this is the uppermost layer of forest floor that is in direct contact with factors influencing the moisture content (relative humidity and air-temperature) it is necessary to assume this for an adequate model. The uppermost litter layer works like a cap for the forest floor layers beneath, so a simplification to assume constant emcs for DMC and DC is justified. Comparing both systems in that sense, the NFDRS is more thorough. A soil layer up to over 10cm is theoretically modelled with variable emcs.

Rain affects the FFMC significantly while there is no real rain effect on the 1 hour moisture code when the observation of the weather (state of the weather) is the same as yesterday's at the point of observation (van Wagner 1975). According to van Wagner (1975) this is justified in open forests where herbs are the main fine fuel and the soils drain quickly. Soils with high duff layer and low rate of drainage though are badly represented.

Furthermore, it is not completely true that the NFDRS models soil conditions. The dead fuels of the NFDRS do not really represent organic layers, since the used parameters (including time-lag) are only derived from dead roundwood fuels in the laboratory. In that sense, all the Dead Fuel Moisture classes are rather comparable to the FFMC but DMC and DC which represent decomposing or decomposed organic matter don't find an equivalent in them. A result of this is that the Moisture classes of the NFDRS don't really account for drought conditions. That's why the KBDI became an obligatory part of later versions of the NFDRS. The KBDI works differently compared to the DC only considering the first 20cm of forest floor and not an 18cm-layer of soil around 9cm below the surface. Nevertheless, it is an index for drought and a major point of criticism by van Wagner (1975) is no more valid.

A point to criticize in the FWI system is that DMC and DC are affected by length of day only as a function of the calendar month (van Wagner 1987). However, latitude is not considered and since the FWI system operates over a range of 20° in latitude this is a factor of uncertainty in the prediction (van Wagner 1987).

Live fuels are totally missing in the FWI in contrast to the NFDRS. Although the jack- or lodgepole pine fuel type has only minor understory vegetation, this is a significant lack in the FWI system. On the other hand, seeing how much input it needs for properly modelling live fuels in the NFDRS this simplification eliminates a huge amount of effort needed to run the system. Also, the fuel moisture codes in the FWI system can model way higher moisture contents than the dead fuel moisture content classes in the

NFDRS. So the NFDRS may underestimate the moisture in the forest floor but adequately represent moisture of live fuels. The FWI on the other hand may adequately estimate the moisture in the soil but neglect moisture in live fuels. Considering this there might be an evening-up-effect in both systems.

In the following the outputs of the two systems will be compared. The ISI (FWI system) is equivalent to the SC (NFDRS). Both predict the potential Rate of Spread without any physical unit. The ISI is only using windspeed and moisture content from the FFMC whereas the SC requires apart from windspeed also the 1-hour Moisture Content, slope and state of vegetation (van Wagner 1975). Slope effects have a crucial impact on the ROS. The FWI system doesn't consider slope effects at all so the NFDRS wins in this point.

Comparable to the BUI (FWI) is the ERC (NFDRS). The ERC considers all moisture classes whereas the BUI only needs DMC and DC since fine fuels (represented by FFMC) are assumed to always burn in a fire and don't contribute to the build-up of a fire (van Wagner 1987; Schlobohm & Brian 2002). A weakness of the NFDRS system is again that the dead fuels are all assumed to be roundwood. Theoretically with the state of the roundwood fuels a build-up effect can be estimated. Nonetheless, the empirical data of the FWI system comes closer to reality.

Both systems rate fire danger through either FWI or BI and indicate a difficulty of control in a general range from Low to Extreme. Both of the system outputs can be used to guide restrictions, determine the level of preparedness, support funding, back requests for resources, facilitate communication and most of all help fire managers make a decision. A surplus of the NFDRS though is that it gives a concrete output of the "staffing level" which means that the system basically tells a fire manager how much of his staff should be on stand-by (Schlobohm & Brian 2002). Nevertheless, it is an easy task to attach this to the FWI output as well.

5. Conclusion

One has to know the limitations and assumptions of a model. This principle always applies in science, so does it here. NFDRS and FWI have both their strengths and weaknesses. Van Wagner (1975) presumes that the NFDRS predicts best in forest common in the US that are open in structure favoring the growth of grasses and brushes and only little of no duff layer. The FWI system in contrast, makes its best predictions in the closed forests of Canada with large litter and duff layer and few variation of herbaceous vegetation throughout the year (van Wagner 1975). Today, this statement is almost 40 years old and especially the NFDRS system has changed since but in essence van Wagner (1975) is still right after the comparison above.

To my eyes, the widespread use of the Canadian Forest Fire Danger Rating System (CFFDRS) (Tian et al. 2005) in which the FWI is a part of, is a proof for the success of this system. The principles of the FWI can be well recommended as a rating of fire danger in countries with one dominating fuel type. All in all, the FWI wins over the NFDRS through its relative simplicity in use and understanding. The few inputs can be provided easily and also cost-effectively. This doesn't mean that the FWI is the sole winner of this comparison. The question remains if more accuracy is needed. In that case, both systems can benefit from each other. A possible synergy of FWI and NFDRS could be a model similar to the FWI system that allows including special site-descriptors similar to the ones of the NFDRS. This would allow a more locally specific prediction of fire danger especially in a country with a significant heterogeneity in fuel types. Also the inclusion of a predictor of live fuel moisture should be taken into consideration.

6. References

Alexander, M. 2008. US BEHAVE System - Wildland Fire Behavior Specialist Course Unit IV-F - powerpoint. Canadian Interagency Forest Fire Centre

Anderson, H.E. 1969. Heat transfer and fire spread. USDA Forest Service, Intermountain Forest and Range Experiment Station, Research Paper INT-RP-69. (Ogden, UT)

Brown, J. H., and M. V. Lomolino. 1998. Biogeography. Sinauer Associates, Sunderland.

Deeming, J. E.; J. W. Lancaster; M. A. Fosberg; R. W. Furman and M.J. Schroeder. 1972. The National Fire-Danger Rating System. USDA Forest Service, Rocky Mountain Forest and Range Experiment Station, Research Paper RM-84, Ft. Collins, Colorado. 165 pp. Revised 1974.

Schlobohm, P. and Brain, J. 2002. Gaining an understanding of the National Fire Danger Rating System. USDA Forest Service. US Department of Interior, National Wildfire Coordinating Group, Fire Danger Working Team, PMS, 932.

Tian, X. R., Mcrae, D. J., Boychuk, D., Jin, J. Z., Shu, L. F., & Wang, M. Y. (2005). Comparisons and assessment of forest fire danger systems. Forestry Studies in China, 7(1), 53-61.

Van Wagner, C. E. 1975. A Comparison of the Canadian and American Forest Fire Danger Rating Systems. Canadian Forestry Service, Petawawa Forest Expt. Stn. Information Report PS-X-59.

Van Wagner, C. E. 1987. Development and structure of the Canadian forest fire weather index system (Vol. 35).

Wotton, B. M.; Alexander, M. E. and Taylor, S. W. 2009. Updates and revisions to the 1992 Canadian forest fire behavior prediction system. Great Lakes Forestry Centre.